# EXPERIMENTS WITH HEAT

## A TRUE BOOK®

by

### Salvatore Tocci

**Children's Press®**
A Division of Scholastic Inc.

New York  Toronto  London  Auckland  Sydney
Mexico City  New Delhi  Hong Kong
Danbury, Connecticut

A metal welder

*Reading Consultant*
**Nanci R. Vargus, Ed.D**
*Primary Multiage Teacher*
*Decatur Township Schools*
*Indianapolis, Indiana*

*Science Consultant*
**Robert Gardner**

*The photo on the cover shows a Bunsen burner being used to heat a liquid. The photo on the title page shows a girl roasting a marshmallow over a campfire.*

**The author and publisher are not responsible for injuries or accidents that occur during or from any experiments. Experiments should be conducted in the presence of or with the help of an adult. Any instructions of the experiments that require the use of sharp, hot, or other unsafe items should be conducted by or with the help of an adult.**

Library of Congress Cataloging-in-Publication Data

Tocci, Salvatore.
Experiments with heat / Salvatore Tocci.
    p. cm. - - (A True Book)
    Includes biographical references and index.
    Summary: Presents experiments that introduce and explain the concept of heat.
    ISBN 0-516-22510-3 (lib. bdg.)    0-516-29365-6 (pbk)
    1. Heat—Experiments—Juvenile literature. [1. Heat—Experiments.
2. Experiments.]   I. Title. II. Series.
QC256 .T63 2002
536'.078—dc21
                                                    2001004937

# Contents

Why Do People Sweat?     5

What Is Heat?     9
  Experiment 1: Moving Molecules
  Experiment 2: Moving Faster

Where Does Heat Come From?     16
  Experiment 3: Cooking a Hot Dog
  Experiment 4: Releasing Heat
  Experiment 5: Conducting Heat

What Is Temperature?     28
  Experiment 6: Taking the Temperature
  Experiment 7: Absorbing Heat
  Experiment 8: Warming Earth

Fun With Heat     41
  Experiment 9: Blowing Up a Balloon

To Find Out More     44

Important Words     46

Index     47

Meet the Author     48

# Why Do People Sweat?

Have you ever been outdoors when the **temperature** was higher than 100°F (37°C)? If you have, then you probably did not feel like doing much except look for some shade. Even in the shade, you may have been sweating because it was still hot. Sweating is the body's way of trying to keep cool.

About 200 years ago, a scientist performed an interesting experiment to show what sweating can do. The scientist took some friends, a dog, and a steak into a room. The temperature inside the room was kept at 126°F (52°C). After 45 minutes, everyone came out of the room. They were all fine, including the dog. The steak, though, was cooked. The people and the dog were able to keep their bodies cool enough—the people by sweating and the dog by panting.

Sweating causes drops of water to collect on the skin. When it's hot, these water drops evaporate from the skin. As the water evaporates, the skin becomes cooler. Normally, a person loses about 1 quart of water in a day through sweating. This must be replaced by drinking water or other liquids.

It is important to replace the water lost through sweating.

If the water is not replaced, then the body cannot sweat. If the body cannot sweat, then it cannot get rid of something that is building up inside—**heat**.

# What Is Heat?

Heat is a type of **energy**. What do you think of when you hear the word energy? Perhaps you think of a person who is very busy or very active. Scientists think of energy as what is needed to make something move or do some type of work. As a type of energy,

These children obviously have a lot of energy.

heat then must be able to
make something move. What
can heat move?

# Moving Molecules

**You will need:**
• several ice cubes

Place the ice cubes in your hands. Notice what happens as heat from your hands warms the ice cubes. It's no surprise that the ice cubes melt and turn into water. How does this simple experiment show that heat is a type of energy?

As you know, an ice cube is just frozen water. Water is made up of very tiny bits, or particles. These particles are called **molecules**.

The water molecules in ice stay pretty much in place. As long as it is cold, these molecules do not have enough energy to move around.

Heat gives these molecules energy. This heat comes from your hand and from the air in the room. Once this heat supplies enough energy, the molecules in ice can move away from each other. When the molecules move far enough apart, ice turns into water. If just a little heat can make molecules move, what happens if you supply even more heat?

Heat supplies the molecules in ice with energy. When they have enough energy, the molecules move far apart. Ice then changes into water.

# Moving Faster

**You will need:**
- two identical glass jars
- refrigerator
- adult helper
- pot
- stove
- oven mitt
- large plate
- food coloring

Fill one jar halfway with water. Place the jar in the refrigerator so that the water gets cold. While the water is cooling, ask an adult to heat some water in the pot. The water should be hot but not boiling. Ask the adult to fill the other jar halfway with the hot water. Use the oven mitt to place the jar with hot water on the plate. Take the jar of water out of the refrigerator and place it next to the one with the hot water. Add two drops of food coloring to each jar, but do not stir the liquid. Watch what happens in both jars.

13

Notice that the food coloring spreads out faster in hot water than it does in cold water.

The hot water has more heat than the cold water. The more heat there is, the more energy the water molecules have. The more energy they have, the faster the molecules move. So molecules of hot water move faster than molecules of cold water.

Because the molecules of hot water are moving faster, they bump into each other harder than molecules of cold water do. Like water, food coloring is made up of molecules. Molecules of hot water also bump into the food coloring molecules harder. This causes the food coloring molecules to spread out faster through the hot water.

The electricity to power all these lights in Seattle, Washington, comes from heat.

You have learned that heat makes molecules move. As a type of energy, heat is also able to do work. Heat is used in power plants to produce electricity. We use this electricity to do many things, such as powering our computers and turning on lights.

# Where Does Heat Come From?

If you have ever been out-doors when the temperature was over 100°F (37°C), then you know how much heat can come from the sun. What else can the heat from the sun do besides make you feel hot?

# Cooking a Hot Dog

**You will need:**
- empty shoe box
- aluminum foil
- nail
- unpainted wire hanger
- liquid soap
- adult helper
- pliers
- scissors
- hot dog

Line the shoe box with aluminum foil. Use the nail to poke a hole in each end of the shoe box slightly below the middle. Wash the hanger with soap and water. Ask an adult to use the pliers to straighten the hanger. Bend the end of the hanger to make a crank with a handle.

Poke the straight end of the wire through one end of the shoe box. Carefully push the

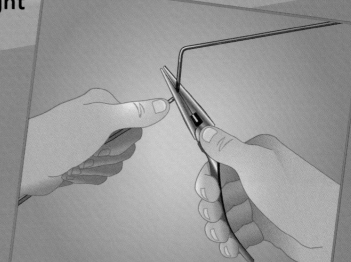

18

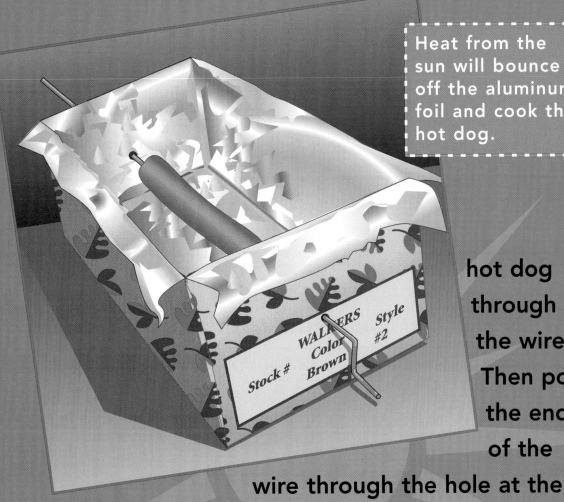

Heat from the sun will bounce off the aluminum foil and cook the hot dog.

hot dog through the wire. Then poke the end of the wire through the hole at the other end of the shoe box. Place your solar cooker in bright sunlight. Every now and then, turn the handle so that the hot dog cooks evenly. Try toasting marshmallows using heat from the sun. Where else can you get heat besides the sun?

WALKERS
Color
Brown

Stock #

Style
#2

19

# Experiment 4
# Releasing Heat

**You will need:**
- large foam cup
- hydrogen peroxide
- plastic spoon
- quick-rising or fast-acting dry yeast

Fill the cup halfway with hydrogen peroxide. Add one spoonful of the yeast to the cup and stir. Watch the contents inside the cup for 5 to 10 minutes. Can you see tiny bubbles of gas coming to the surface? How does the bottom of the cup feel?

FAST-RISING
**YEAST**

Notice the tiny gas bubbles that rise to the surface.

Yeast breaks down hydrogen peroxide to make new substances. When new substances are made, a **chemical reaction** has taken place. One of the substances made in this chemical reaction is oxygen gas. The tiny bubbles you saw were filled with oxygen. Besides making new substances, chemical reactions can also give off heat. Did the bottom of the cup feel warm? This happened because heat was given off by this chemical reaction.

The torch this man is holding uses a chemical reaction that gives off enough heat to weld metal pieces together.

Most power plants get the heat they need to make electricity from chemical reactions. These power plants burn fuels, such as oil and coal. Burning is a chemical reaction that gives off heat.

Burning oil in this California power plant produces steam. This steam spins giant metal blades to produce electricity.

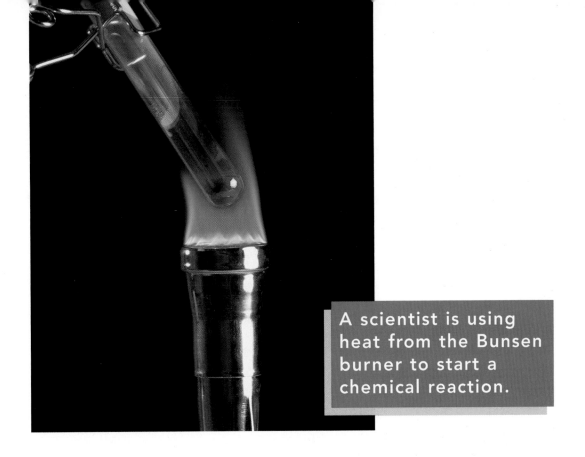

A scientist is using heat from the Bunsen burner to start a chemical reaction.

Not all chemical reactions, however, give off heat. Some chemical reactions take in heat. Besides being given off or taken in, what else can happen to heat?

# Experiment 5

## Conducting Heat

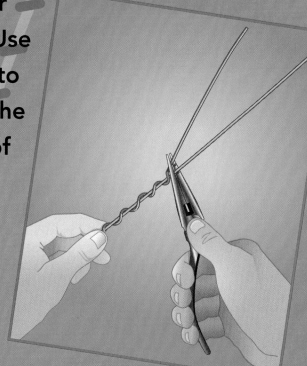

**You will need:**
- adult helper
- wire cutters
- tape measure
- metal coat hanger
- sandpaper
- thick copper wire (used by electricians)
- cutting board
- aluminum foil
- paper clips
- candle
- matches
- pair of oven mitts

**A**sk an adult to help you. Cut a 12-in. (30 cm) piece of coat hanger. Sand off any paint or coating on the wire, which is made of iron. Cut a 12-in. (30 cm) piece of copper wire. Use pliers to twist the ends of the iron and copper wires together.

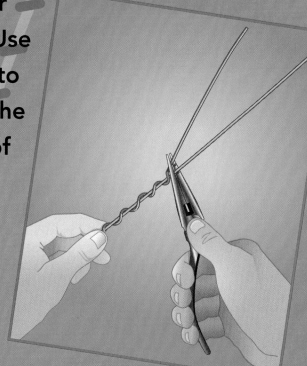

Cover the cutting board with aluminium foil. Place the wires on the foil. Place three paper clips on the iron wire and three paper clips on the copper wire. Ask an adult to light the candle. Hold the candle so that the wax drips onto the paper clips and over the wires. As the wax cools, the paper clips should stick to the wires.

Space out the paper clips evenly. Be careful not to get any hot wax on your hand.

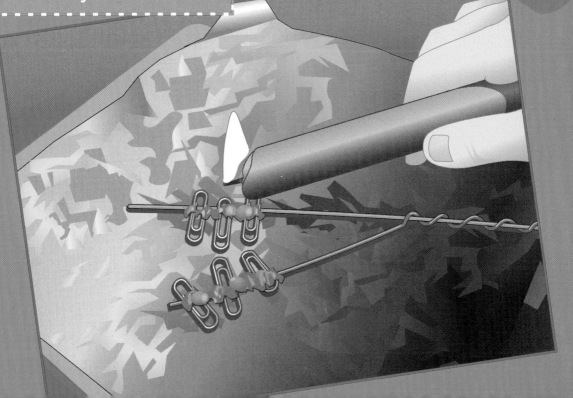

The oven mitts will keep your fingers from getting warm.

Use the oven mitts to hold the wires over the candle flame. Watch what happens to the paper clips. Heat from the candle flame passes along the wires. As it does, the wax melts causing the paper clips to drop. Which paper clips fall off first?

The paper clips should start falling off the copper wire first. This happens because heat travels faster through copper than it does through iron. The heat reaches the wax on the copper wire sooner than it reaches the wax on the iron wire. The wax melts sooner on the copper wire, allowing the paper clips to fall. Copper is said to be a better **conductor** of heat than iron.

# What Is Temperature?

When you go outdoors on a sunny day in the summer, you can tell that it's hot. You need a thermometer, however, to tell exactly how hot it is. A thermometer tells you the temperature. Temperature is a measurement that shows how hot or how cold something is.

This girl knows how hot and sweaty she can get playing tennis on a summer day.

Do you really need to know the temperature to tell if something is hot or cold?

# Experiment 6

# Taking the Temperature

**You will need:**
- various materials with one smooth surface larger than the size of your hand (wood, cardboard, glass, metal, plastic)
- liquid crystal thermometer card (available at a pet store)
- pencil
- paper

**S**et all the materials so that they are not in direct sunlight or near a light source. Place your palm flat on the surface of each material. Which ones feel warm? Which ones feel cool? Arrange the materials in order from warmest to coolest. Now place the thermometer on the material that felt the warmest. Wait until the temperature no

Does the
surface
feel cool
or warm?

longer
changes. Write down the
temperature of the material that felt the warmest.
Do the same with the other materials. What do you
notice about the temperatures of all the materials?

Were you surprised to find that all the materials have the same temperature? So why did some
materials feel warm while others felt cool?

31

Metals feel cool. Remember that metals, such as copper and iron, conduct heat. When you touch metal, it conducts heat away from your hand. Heat keeps flowing from your hand to the metal. Because heat keeps leaving your hand, the metal feels cool when you touch it.

Even though it has the same temperature, wood feels warmer than metal. Unlike metal, wood does not conduct heat. When you touch it, very little heat flows from your hand to the wood. Because very little heat leaves your hand, the wood does not feel as cool as the metal.

All these materials will have the same temperature if left in the same place.

On a hot summer day, you may have gone to a beach near the ocean or a large lake. Even before you went into the water, you felt cooler than you did at home. Why is it always cooler near the water in the summertime?

# Experiment 7

# Absorbing Heat

**You will need:**
- two large foam cups
- water
- soil or sand
- lamp with incandescent bulb (optional)
- thermometer

**F**ill one cup with water and the other with soil or sand. Place both cups in direct sunlight or under a lamp for several hours. As the water and soil take in heat, their temperatures should rise. Measure the temperature just beneath the surface of the water. Then, measure the temperature just beneath the surface of the soil. Which is warmer?

34

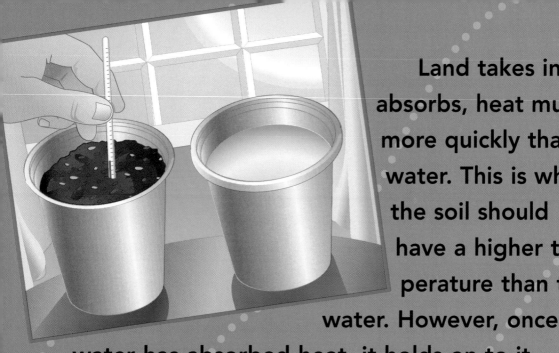

Land takes in, or absorbs, heat much more quickly than water. This is why the soil should have a higher temperature than the water. However, once water has absorbed heat, it holds on to it longer than land does. In other words, water does not give up heat easily. So the air near water does not get as hot as the air in inland areas. This is why it's always cooler at a beach on a hot day. How hot can it get?

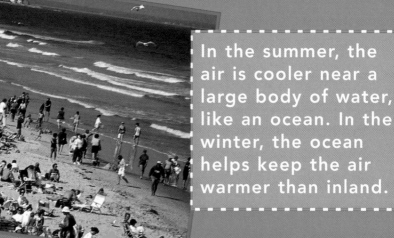

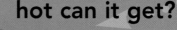

In the summer, the air is cooler near a large body of water, like an ocean. In the winter, the ocean helps keep the air warmer than inland.

# Experiment 8

# Warming Earth

**You will need:**
- measuring cup
- two identical, small glass jars
- plastic bag
- twist tie
- lamp with incandescent bulb (optional)
- thermometer

Pour the same amount of cold water into each jar. Use the thermometer to make sure that the temperature of the water in both jars is the same. Wrap one jar in a plastic bag. Seal the bag with a twist tie. Place both jars in direct sunlight or under a lamp for two hours.

If you have ever walked on asphalt in the summer, you know how hot the sun can make the ground get.

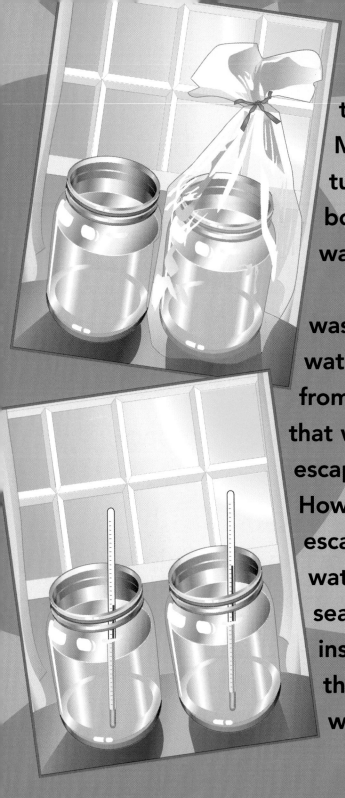

Remove the jar from the plastic bag. Measure the temperature of the water in both jars. Which is warmer?

Heat from the sun was absorbed by the water in both jars. Heat from the water in the jar that was not sealed could escape back into the air. However, heat could not escape as easily from the water in the jar that was sealed. Heat then stayed inside the bag, making the water warmer.

37

The same thing may be happening to Earth. Homes, industries, and cars have been burning more and more fuels. All of this burning releases gases into the air. These gases collect in the atmosphere. These gases act like the bag that covered the jar. They keep in the heat that Earth receives from the sun. This heat may be making Earth warmer than it should be. This is called **global warming**.

Notice that the arrows that show heat are curving back to Earth because heat is trapped by the gases that collect in the atmosphere.

Heat is a type of energy that can be released, absorbed, conducted, and measured. Heat is used by power plants to produce electricity. On hot days, heat may build up inside our bodies. To get rid of it, we sweat. Heat may also be building up on Earth. If we cannot get rid of this heat, then Earth may get warmer and warmer. Scientists are not sure how this could affect life on Earth.

# Fun With Heat

You learned that water can absorb heat from the sun. Water can also absorb heat from a candle flame. See how you can use this fact to teach your family and friends something about heat.

# Blowing Up a Balloon

**You will need:**
- two large, round balloons,
- measuring cup
- water
- adult helper
- matches
- candle

**B**low up one balloon and tie it closed. Pour a quarter cup of water into the other balloon. Then inflate the balloon and tie it closed. Ask an adult to light a candle. While you hold the balloon in the air, have the adult hold the candle under the balloon. The flame should touch the balloon. What happens to the balloon?

Now have the adult hold the candle so that the flame is right under the water in the other balloon. What happens to this balloon?

The balloon with no water
should burst. Heat warms the balloon and
weakens it. The balloon becomes so weak that
the air inside breaks open the balloon. The
balloon with water should not burst. The water
absorbs the heat from the flame. The balloon
remains strong enough to keep the air inside.

# To Find Out More

If you would like to learn more about heat, check out these additional resources.

 **Books**

Gardner, Robert and Eric Kerner. **Science Projects about Temperature and Heat.** Enslow Publisher, 1994.

Gordon, Maria and Mike Gordon. **Fun With Heat.** Thomson Learning, 1995.

Halacy, Beth and Dan Halacy. **Cooking With the Sun: How to Build and Use Solar Cookers.** Morning Sun Press, 1992.

Lauw, Darlene and Lim Cheng Puay. **Heat (Science Comes Alive).** Crabtree Publishers, 2001.

Searle, Bobbi. **Heat and Energy (Fascinating Science Projects).** Copper Beach Books, 2001.

Wood, Robert. **Physics for Kids: 49 Easy Experiments With Heat.** Scholastic, 1990.

# Organizations and Online Sites

**Arizona-Sonora Desert Museum**
2021 N. Kinney Road
Tucson, AZ 85743-8918
520-883-1380
*http://www.desert museum.org/index.html*

Click on "Desert Life" to learn how plants and animals manage to survive the extreme heat of the desert.

**Environmental Protection Agency (EPA)**
1200 Pennsylvania Avenue, NW
Washington, DC 20460
*http://www.epa.gov/ globalwarming/kids/index. html*

The EPA is the U.S. government agency in charge of protecting our environment. Log on to their site and click on "Kids & Educators." You will be led to activities, games, and links that deal with global warming.

**The Exploratorium**
3601 Lyon Street
San Francisco, CA 94123
415-EXPLORE
*http://www.exploratorium. edu/snacks/snackintro.html*

This site has a list of "snacks," which are short activities that you can do at home. Some of these "snacks" involve experiments with heat.

**Reeko's Mad Science Lab**
*http://www.spartechsoftware. com/reeko/Experiments/Exp SteelWoolGenerating Heat.htm*

Find out how to cause a chemical reaction between steel wool and vinegar that releases heat. You will also learn more about what goes on in a chemical reaction that releases heat.

# Important Words

*chemical reaction* a process that makes new substances and can either release or absorb heat

*conductor* something, like a metal, that can pass heat along

*energy* what is needed to move something or do some type of work

*global warming* slow heating of Earth caused by gases that collect in the atmosphere

*heat* a type of energy

*molecule* an extremely small particle that makes up water and many other substances

*temperature* a measurement of heat

# Index

(**Boldface** page numbers indicate illustrations.)

absorbing heat, 34–35, **36,** 40, 41, 42–43
atmosphere, 38, 39
balloon, 42–43
beach, 33, **33,** 35
blowing up a balloon, 42–43
bubbles, 20–21
burning, 22, 38
chemical reaction, 21, 22, 23
cold, 28, 29
conducting heat, 24–27, 32, 40
cooking, 18–19
dog, 6-7
drinking, 7, **8**
Earth, 36–37, 38, 39, 40
electricity, 15, 22, **22,** 40
energy, 9, **10,** 11, 12, 14, 15, 40
evaporation, 7
fuel, 22, 38
gas, 20–21, 38, 39
global warming, 38–39

heat, 8, 9–10, 11, 12, 14, 15, 16, 19, 21, 22, 23, 39, 40
  absorbing, 34–35, **36,** 40, 41, 42–43
  conducting, 24–27, 32, 40
  measuring, 28, 30–32, 40
  releasing, 20–21, 40
hydrogen peroxide, 20, 21
ice, 11, 12
measuring heat, 28, 30–32, 40
melting, 11
metals, 21, 32
molecules, 11–12, 14, 15
moving, 10, 11–12, 13–14
oxygen, 21
panting, 7
power plants, 22, **22,** 40
releasing heat, 20–21, 40
skin, 7
solar cooker, 19
sun, 16, **17,** 19, 36, 41
sweating, 5–8, 40
temperature, 5–628–29, 30–32
water, 7, 8, **8,** 11, 12, 13, 14, 33, 41, 43
yeast, 20, 21

47

# Meet the Author

Salvatore Tocci is a science writer who lives in East Hampton, New York, with his wife, Patti. He was a high school biology and chemistry teacher for almost thirty years. As a teacher, he always encouraged his students to perform experiments in order to learn about science. During the warmer months, he gets relief from the heat by sailing on the water in his sail boat.

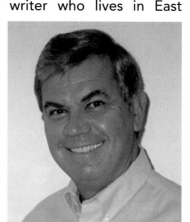

Photographs © 2002: Dembinsky Photo Assoc.: 4 (Mark E. Gibson), 29 (Ron B. Smith); EyeWire/Getty Images: 1; International Stock Photo: 15 (Chad Ehlers), 36 (Johnny Stockshooter); Photo Researchers, NY: 17 (Van Bucher), 33 (Christopher Marona), 12, 21 (Charles D. Winters); PhotoEdit: 8 (Amy Etra), 22, 23 (Tony Freeman); Rigoberto Quinteros: 11; Stone/Getty Images: cover; The Image Bank/Getty Images: 2; The Image Works/Kathy McLaughlin: 10; Visuals Unlimited/Jeff Greenberg: 35 bottom.
Illustrations by Patricia Rasch